I0469319

Industrial Silo Fire and Explosion
Charlotte, North Carolina

Reported by: John Kimball

This is Report 122 of the Major Fires Investigation Project conducted by Varley-Campbell and Associates, Inc./TriData Corporation under contract EME-97-C0-0506 to the United States Fire Administration, Federal Emergency Management Agency.

Department of Homeland Security
United States Fire Administration
National Fire Data Center

U.S. Fire Administration Fire Investigations Program

The U.S. Fire Administration develops reports on selected major fires throughout the country. The fires usually involve multiple deaths or a large loss of property. But the primary criterion for deciding to do a report is whether it will result in significant "lessons learned." In some cases these lessons bring to light new knowledge about fire--the effect of building construction or contents, human behavior in fire, etc. In other cases, the lessons are not new but are serious enough to highlight once again, with yet another fire tragedy report. In some cases, special reports are developed to discuss events, drills, or new technologies which are of interest to the fire service.

The reports are sent to fire magazines and are distributed at National and Regional fire meetings. The International Association of Fire Chiefs assists the USFA in disseminating the findings throughout the fire service. On a continuing basis the reports are available on request from the USFA; announcements of their availability are published widely in fire journals and newsletters.

This body of work provides detailed information on the nature of the fire problem for policymakers who must decide on allocations of resources between fire and other pressing problems, and within the fire service to improve codes and code enforcement, training, public fire education, building technology, and other related areas.

The Fire Administration, which has no regulatory authority, sends an experienced fire investigator into a community after a major incident only after having conferred with the local fire authorities to insure that the assistance and presence of the USFA would be supportive and would in no way interfere with any review of the incident they are themselves conducting. The intent is not to arrive during the event or even immediately after, but rather after the dust settles, so that a complete and objective review of all the important aspects of the incident can be made. Local authorities review the USFA's report while it is in draft. The USFA investigator or team is available to local authorities should they wish to request technical assistance for their own investigation.

For additional copies of this report write to the U.S. Fire Administration, 16825 South Seton Avenue, Emmitsburg, Maryland 21727. The report is available on the Administration's Web site at http:// www.usfa.dhs.gov/

U.S. Fire Administration

Mission Statement

As an entity of the Department of Homeland Security, the mission of the USFA is to reduce life and economic losses due to fire and related emergencies, through leadership, advocacy, coordination, and support. We serve the Nation independently, in coordination with other Federal agencies, and in partnership with fire protection and emergency service communities. With a commitment to excellence, we provide public education, training, technology, and data initiatives.

ACKNOWLEDGMENTS

The U.S. Fire Administration greatly appreciates the cooperation received from the following people and organizations during the preparation of this report:

C.A. "Chuck" Gallyon, II Fire Marshal
Iredell County Fire Department
Statesville, North Carolina

Tim Fox Assistant Fire Marshal
Iredell County Fire Department
Statesville, North Carolina

Chief Wayne Mayberry Monticello Volunteer Fire Department
Statesville, North Carolina

Mark Mason Rescue Chief
Troutman Volunteer Fire Department

Daniel Cline Deputy Chief
Statesville Fire Department

Captain Calvin Stovall Statesville Fire Department

Larry Dickerson Director
Iredell County Emergency Management

TABLE OF CONTENTS

Industrial Silo Fire and Explosion
Charlotte, North Carolina
December 1997

OVERVIEW

On December 21, 1997, three volunteer firefighters from Iredell County, North Carolina were injured in an explosion in a silo used to collect and store wood waste for utilization as fuel at a cord reel manufacturing facility. The silo was an agricultural type that had been converted for use as a collector for sawdust. The structure had been the site of a minor explosion five years previously that had caused no injury.

The firefighters had been directing water into the silo for over two hours from opening in the silo roof when the decision was made to access the wood product inside. A loud, low order explosion destroyed the top of the silo and endangered the firefighters who had been operating on the roof. The explosion buffeted personnel operating on the ground as well.

One of the three firefighters operating on the roof was lifted upward and landed back in the silo, his call cushioned by the fill product. Another was ejected up and outward. He fell through a trailer shed and landed in an open top trailer filled with wood product. The third was enveloped by the machinery from the roof top which trapped him at the top rim of the silo. All three were rescued in the course of a multi-hour operation. The firefighter who landed in the silo was treated for burns and released from the hospital a week later. The firefighter who landed in the trailer suffered shoulder and knee injuries requiring surgery, and the one trapped at the top rim of the silo was treated and released for minor burns and bruises.

This incident highlights the need for the **recognition of the dangers of oxygen-limiting silos** regardless of their use and setting. Other issues identified are the need for a hazard and risk assessment process in decisionmaking on the fireground, the importance of site control and accountability, the need for group training in technical rescue operations, the coordination of non-fire department resources and the role of emergency management personnel fulfilling an active role in a unified command structure.

SUMMARY OF KEY ISSUES

Issue	Comment
Recognition of oxygen-limiting silos	Oxygen limiting silos pose explosive hazards regardless of their use and application. In agricultural settings, the recommended action for fires of this type is to close off silo and allow fire to burn itself out.
Risk assessment	Incident commanders should very carefully evaluate whether direct attack or defensive tactics are appropriate to the risk involved. If the decision is made to attack, operations should be conducted from the top using remote devices or penetrating nozzles.
Accountability	All personnel should be subject to an accountability system with the safety officers monitoring the activities and whereabouts of all personnel on site.
Communications	Little information was relayed to Communications after the explosion. Separate radio channels were used by different jurisdictions and organization, all of which contributed to confusion and complicated response.
Use of aerial device	Elevated streams such as ladder pipes, tower ladders, or telesquirts could have been utilized to safely apply water instead of placing firefighters at risk.
Use of protective clothing	The use of full turnout gear prevented more serious injury.
Use of harnesses	The deployment of Class III harnesses on the attack crews facilitated rescue after the explosion and eliminated the need for affixing additional knots or belts to rescue the injured firefighters. The anchor points were the weak links, however.
Technical rescue training	Technical rescue crews had good basic skill level. However, this incident highlighted the need for the technical rescue trained personnel to train together as a unit, regardless of departmental affiliation.
Water Supply	Tanker shuttles are adequate for handlines; sustained flow of master streams for flooding operations indicates large diameter supply lines from hydrants.

BACKGROUND

Iredell County, North Carolina

Iredell County, North Carolina, is located just north of Charlotte, NC, with a land area of 574 square miles and a population of nearly 100,000. The county seat, Statesville, is located about 40 miles north of Charlotte and has a population of about twenty-one thousand people. The county is bisected by Interstate routes 77 and 40 which intersect very near the city of Statesville. The geography is rolling hills with no extremes of elevation or other prominent features. The county is primarily rural and agriculture is the major source of employment. The industry is mixed with a variety of light manufacturing, electronics and transportation-related services. Furniture manufacturing and other wood-related products are prominent industries in Iredell County.

Emergency Management

The county is served by an Emergency Management Director who functions as liaison among the various emergency services, including police, fire, and emergency medical response. The incident management role of the Emergency Management Director is limited to disasters declared by the governor. On other emergency incidents he functions as a resource coordinator in a unified command system, supporting the tactical response functions. One of the key duties of the Emergency Management Director is the operation and supervision of the Emergency Communications Center which receive and processes all calls for emergency services.

On significant incidents, Communications alerts the Emergency Management Director, Fire Marshal, Sheriff and the Emergency Medical Services Director. These managers respond to the scene, proceed to the Emergency Operation Center, or monitor the situation, as appropriate, to coordinate response and serve as liaison with other agencies.

Fire Protection

Fire protection is provided by a network of volunteer fire departments that are supported by the Iredell County Fire Marshal's Office. The role of the fire marshal is to assist the county volunteer fire departments in fire protection activities, perform public fire prevention education, and complete fire inspections in commercial and public buildings. The fire marshal is also tasked with assisting the county volunteer departments in firefighting activities, and is responsible for fire cause and origin investigations as well as the coordination of investigation efforts with local and State law enforcement officials.

The Iredell County volunteer fire departments operate from 21 fire stations staffed by 500 certified fire personnel and housing 102 pieces of equipment. The City of Statesville and the Town of Mooresville operate fully career departments.

Emergency Medical Services

Emergency medical service is provided by a network of fully paid advanced life support units and volunteer rescue squads which supply basic life support services. The volunteer rescue squads are either a part of the volunteer fire departments or operate as independent organizations. All emergency medical services, however, operate under the authority of the County Medical Director. The fully staffed advanced life support organization maintains a minimum of six fully staffed ALS units operating from four bases located in the county.

Aeromedical service is provided by eight medevac services responding from Regional trauma centers. The three closest services to Iredell County operate three helicopters from Charlotte, two from Winston-Salem, and one from Durham.

Facility

The manufacturing facility is one of two owned by a company that produces wooden reels that house cable, cord, rope, and other flexible products. The main building is approximately 300' by 400' having a square footage of 120,000. Located just outside the city limits of Statesville in Iredell County, the building is of non-combustible construction with a metal roof covered with tar and gravel over flat metal decking and steel trusswork.

The silo was used for the collection, retention, and distribution of wood waste from the manufacturing process. The wood waste consists of shreds, shavings, and sawdust collected from a ductwork system in the plant via a cyclone and ducting drawn to the top of the silo. Another cyclone on top of the silo deposits the product into the silo.

The top configuration of the silo consisted of the integral dome shaped roof, (part of the original manufactured structure), and the cyclone (a fan and collector structure for the movement and direction of the wood product). The cyclone and housing were supported by a framework of steel I-beams. A steel catwalk around the rim of the silo was accessed by two separate steel-cage ladders. As the wood waste was collected and stored, it was removed from the bottom by a means of an electric-powered double shaft screw auger to a chute in the bottom of the silo to the facility boiler system. The North Carolina State Department of Environment and Natural Resources issued the permit for this installation.

Photo #1 gives a perspective of the silo and its orientation to the building. The trailer shed to the right received wood waste via ducting and cyclone arrangement similar to that of the silo, for short-term storage in an open-box trailer and then disposal off site.

The silo was an oxygen-limiting type designed to store food for livestock. The design permitted the silage to be stored for long periods by preventing the oxygen in the air to penetrate the structure. By their nature, oxygen-limiting silos are very strong structures built to exclude air exchange with the ambient atmosphere outside the silo. The shell of this unit lacked total integrity because it was compromised at the two access panels. These panels were not part of the original configuration that featured airtight coating over the bolts in the interior. The access panel bolts lacked this coating.

In an agricultural setting, this type of silo has been the cause of fatal explosions, which have taken the lives of firefighters, (8/27/85, Marshallville OH, 3 firefighters killed, 8/5/93, Morgan County GA, 2 firefighters killed.) The mechanism of this type of explosion is a backdraft-like phenomenon caused when oxygen-containing air is introduced into areas of built-up heat and gases in deep-seated, slow-burning areas of the stored agricultural products. All that is needed for a violent combustion explosion is a source of oxygen. The presence of wood dust could likely have contributed to and aggravated the explosion.

In an agricultural setting, with the fuel source being finely chopped organic material (silage), the commonly recommended tactic is to close off all silo openings, assure there is no fire outside the silo structure, and close off the silo and allow the fire to burn itself out. A factory-installed connection permits the introduction of fire-inerting nitrogen to blanket the fire in the interior. This is the only recommended extinguishing action of an oxygen-limiting silo in an agricultural setting.

Location History

Over the course of approximately ten years, the fire department responded to at least thirteen fire alarms at this facility. The silo itself was the scene of a previous fire and explosion on July 29, 1993. In that incident the explosion bowed the roof upward into a mushroom configuration. No one was injured in that explosion and damage from the incident was minor.

Incident Narrative

On Sunday, December 21, 1997, units from Monticello, Troutman, and West Iredell Volunteer Fire Departments responded with six pumpers and two chief officers to the manufacturing facility at 1475 Winston Avenue for a report of fire in a silo. The call was dispatched at 0858 and the first unit responded at 0901. Weather conditions were overcast with no precipitation and the temperature was in the high 30s to low 40s. There was no appreciable wind.

The first unit on the scene (from Monticello), reported moderate smoke coming from the top of the silo. The Chief of the Monticello VFD established command and proceeded to investigate fire conditions. The fire was confined to the silo with no extension to the manufacturing facility and no fire travel in the ductwork. Command decided that the best course of attack would be a flooding operation, using handlines, from the top of the silo. The strategy was to maintain the steady flow of an 1¾" attack line set on 30-degree fog to penetrate and extinguish the burning wood material.

The decision also was made to utilize a water shuttle rather than to deploy a supply line from the hydrant at the facility's driveway entrance. The distance to the hydrant was approximately 2,000 feet. (This decision limited the water flow potential in terms of gallons per minute.) Teams of firefighters worked from the roof area of the silo, directing the hose stream through openings in the silo roof to penetrate the burning material. Personnel were fitted with Class III harnesses and anchored in with 2" webbing to tie-off points on the silo's catwalk railing.

The flooding operation continued for approximately 2½ hours and flowed an estimated 10,000 gallons of water into the silo. Support personnel responded with air units, standby ambulances, and rehab supplies. Firefighters were rotated through working assignments and break times.

The smoke had receded to a very light condition when the decision was made to transition the operation from attack to overhaul and complete the extinguishment of any remaining embers. Firefighters were to manually remove the contents rather than enter the silo. Command decided to remove the two access plates on the bottom of the silo to access the contents. One of these plates was approximately 2 feet x 2 feet; the other was 3 feet x 4 feet. All indications were that no hot spots remained, and the general exterior silo area was cool.

The Chief had moved all but essential personnel from around the base of the silo area as the crew commenced to remove the plates. As they worked on removing the last bolt of the 3 x 4 plate, one of the firefighters at the plate area noticed water flowing out along with burning ashes in the air space above. He then described a sucking and gurgling sound that followed by a loud, low order explosion. The firefighters on the roof recalled the same type of inrush of air at their position followed by a dull thump and the roar of the main explosion. (Photo #2 shows access plate in relation to silo, photo #3 is a closer view of the access plate area.)

Eight firefighters were near the access plate area at the base of the silo. Several were knocked off their feet and some were actually propelled through the air. Overall, approximately thirty-five people (firefighters, EMTs, and support personnel) were near the silo. Several large pieces of machinery, structural members and roof sections fell to the ground, narrowly missing several firefighters.

The three firefighters on the silo were propelled in different directions. Their life belts were intact, but the webbing and the anchor points (the catwalk) were shredded. One firefighter traveled straight up and landed back into the silo. Another was ejected upward and outward traveling 32 feet landing on the roof of the trailer shed, crashing through the shed roof and into an open top trailer of wood products. The third attack crewman was trapped on the top of the silo by the metal cover of the cyclone that had been a part of the roof-top wood distribution system and the ladder itself.

The incident commander immediately reported an explosion to Communications and requested assistance in the form of medic units and a tower ladder. No other "situation report" information was passed on from the scene for the duration of the incident. Iredell County medic units and the aerial tower from the City of Statesville were dispatched. At this point, because fireground units were operating on a different frequency than were Communications and Statesville, and detailed information was not passed on from the scene. Neither Communications nor assisting units had a clear picture of the conditions at the scene. The two Communications personnel now became overloaded due to the instantaneous increase of radio traffic, dispatch duties and telephone notifications.

The scene was transformed from one of a routine overhaul operation, to one of stunned confusion. What had been a mop-up and demobilization instantly became a technical rescue with many unanswered questions as to the whereabouts and condition of fellow firefighters. Unit 118, the Chief of the Troutman Rescue Squad, reported to Command and worked with the IC to establish a course of action. Unit 118 was assigned the Rescue Sector and he immediately placed the medevac helicopter system on standby.

Initial assessment from the ground indicated three firefighters immediately affected by the explosion and little or no fire remaining in the silo. The roof had blown into several large pieces and the cyclone collector housing at the roof of the silo rested precariously over the trapped firefighter's position.

The immediate priority was accounting for all personnel, locating any who were missing, and assessing the number and severity of injuries. The next course of action was the technical rescue as well as treatment and transport of the injured. The Rescue Sector was assigned the coordination of this phase of the operation.

The first firefighter to be located, accessed, treated, and transported was the member who had been blown into the trailer shed. He was immobilized, removed from the trailer, and transported by ground to Iredell Memorial Hospital at 12:30 PM.

A firefighter was assigned to access the silo roof via the undamaged staircase for a situation report. His investigation confirmed that one member was in the silo and one firefighter was trapped by machinery at the roof line.

The rescue operation at this point was divided into a dual strategy; removing the firefighter in the silo while protecting the other firefighter trapped at the top. The firefighter in the silo was assigned immediate priority because of concern of his condition and fear over the fire condition in the silo (e.g. the possibility of rekindle or subsequent explosion). The firefighter at the roof was in stable condition with no major injuries, and appeared to be only trapped in an enveloping fashion by the machinery.

The aerial tower device from Statesville Fire Department responded and their captain was assigned responsibility for the physical rescue of the firefighters on and in the silo. The vehicle set up in the immediate area, but had to reposition because of obstructions at the silo base and the severe angle and reach required.

One member from Statesville Fire Department rappelled into the silo, assessed the victim, and directed his removal by affixing a lifeline to the injured firefighter intact rescue harness. A haul system was rigged to lift the injured firefighter out of the silo. This effort took several tries due to a lack of standard evolutions and no previous **inter-departmental** training on the part of the technical rescue personnel. The victim was removed and transported at 1:00 PM by air ambulance to North Carolina Baptist Hospital Burn Unit in Winston Salem.

The County Fire Marshal had stopped at a nearby steel fabrication yard enroute to the scene, and requested the services of a mobile heavy-duty construction crane. The construction crane was placed in position to stabilize the structural debris and prevent it from falling further onto the enveloped firefighter. As the damaged machinery was first stabilized, the cyclone and its housing were slowly and carefully lifted to allow the trapped firefighter to extricate himself from his turnout gear and out of his entrapment. He was removed at 3:00 PM and transported via helicopter to NC Baptist Hospital.

While these extrication actions were underway, the County Emergency Manager had reported to the Incident Commander at 1300 hours. The Emergency Manager was instrumental in coordinating non-fire/rescue resources and interacting with the various media representatives. His assistance enabled more efficient use of available resources, and allowed the fire/rescue commanders to concentrate on the extrication and treatment activities.

Investigation

The investigation as to the cause of the fire and subsequent explosion commenced immediately after the scene was released to the fire marshal's office. The scene was deemed to be unsafe for immediate, detailed, investigation due to the structural damage and the heavy structural material around the top of the silo. This damaged material was removed by a contractor in the presence of the investigating fire marshal and the site was then rendered safe for preliminary investigation efforts. Due to worsening weather conditions, the in-depth investigation was deferred and the effort confined to the perimeter area.

The removal of the remaining product in the silo was required before investigators could enter the silo and proceed with the comprehensive investigation. A contractor used an industrial vacuum and removed the remaining product in the silo on the following Monday. The condition of the remaining sawdust and the burn level in the silo was observed. The scene was photographed and the site was documented by drawings. During the first several days after the explosion, the investigation was hampered adverse weather conditions: ice, snow, and cold temperatures.

The observation indicated a large void area (approximately 12 feet by 16 feet by 10 feet) at the bottom of the silo radiating from the screw auger area. One of the "fingers" (Photo #4) of burned material ran to the direction of the access panel that was the focal point of the explosion. The direction of the burned void indicated fire travel in the direction of the screw auger to the compromised areas of the silo shell: the access panels. (Photos #5 and #6.) Despite the application of several thousand gallons of water, **apparently no water penetrated to sufficient depth to reach the burning wood material.**

The fire marshal determined that the fire started when the gear box seized – burning the belt of the drive unit. The actual fire origin was deemed to have come from the plant side rather than from inside the silo. The day of fire origin was estimated to have been Tuesday or Wednesday. It was surmised that the fire burned void areas out of the sawdust over the auger. The fire also burned from the center toward the bolt holes in the access plates. (Photo #7).

The explosion occurred when the 3 foot by 4 foot side panel was removed, thus allowing oxygen to enter the burning area. (Photo #8). Because of the compromise of the integrity of the shell by the bolts and access plate, a minute source of oxygen-bearing air permitted the fire to burn in finger-like fashion from the source in the gearbox to the silo wall. This was an area of deep-seated, incomplete combustion, which needed only a source of oxygen to ignite explosively. This is consistent with backdraft-like explosions that have occurred in agricultural product silos. The sawdust also entrained dust to combine with the rapid oxidation of the backdraft to produce the energy sufficient to damage the silo and injure firefighters. The water applied in the extinguishment attempts apparently either ran off the surface or was absorbed and acted as a tamper over the burned out voids in the wood waste.

LESSONS LEARNED

1. **Backdraft –life explosions in oxygen-limiting silos can occur wherever the product is capable of burning and the air is limited.**

    In an agricultural setting, this type of silo has been the site of explosions which have taken the lives of firefighters. The mechanism of these explosions are a backdraft-like phenomenon, which is the result of oxygen-containing air being introduced into areas built-up of heat and gases from incomplete combustion needing only a source of oxygen for a violent combustion explosion. The explosion was possibly aggravated by the presence of wood dust, which may have added to the fuel load in the form of finely dispersed combustible particles.

    Oxygen-limiting silos pose explosive hazards regardless of their use and application. In deep-seated fires of shredded combustible material, regardless if it is silage, wood chips or dust, the potential exists for fires to reach smoldering stage of high heat and low oxygen. Backdraft-like explosions have occurred in silos that have been converted to conventional-type agricultural silos because of the strong integrity of seams and joints in the shell, which are designed to prevent air incursion.

2. **Hazard and risk assessment procedures should be completed to determine the best attack strategy as to either offensive or defensive mode as well as placement of personnel and attack hose streams.**

    Incident commanders need to carefully evaluate the hazard and risk involved. Defensive tactics or direct aggressive attack may be indicated in cases of high life safety hazards, high property-loss potential, or extreme potential for fire extension. If the fire is contained, or is not threatening life or high property loss, measures to greatly minimize the risk to firefighters should be utilized. Unstaffed hoselines, standby operations, or other low risk-level tactics could be considered.

3. **There were several gaps in the chain of communications between the incident scene, communications, and assisting units. This hampered the effectiveness of rescue operations.**

    Fireground units were on non-repeating tactical channels not commonly monitored by Communications. The Statesville Fire Department operates on a separate radio channel so critical information from the scene was not available. Periodic situation updates have provided during the operation prior to the explosion would have been beneficial in providing background information to Communications. More detailed situation reports from the scene could have been reported to Communications after the explosion. Important information could have been relayed to units responding to assist in the rescue effort. This information could have given assisting units a clear picture of the number of injured, severity of injuries and means of entrapment.

4. **Ladder pipes, tower ladders, or telesquirts could have been utilized to safely apply water with less risk to personnel.**

    In a case where the risk assessment indicates low life or property loss potential, and the structure presents a possibility of explosion, remotely operated devices could be used to safely apply large quantities of water to complete a flooding and penetrating extinguishment effort. The strategic objective of prevention of fire extension to exposures can be safely accomplished with remotely operated master streams.

5. **A well-defined command structure with a safety officer and accountability of all personnel should be in place at all fires, even into the overhaul and demobilization stage. All personnel should be subject to an accountability system with the safety officers monitoring the activities and whereabouts of all personnel on the site. In addition to supervising accountability of personnel, the safety officer can play a major role in the risk assessment process.**

At the time of the explosion, there was no safety officer in the command structure to account for all personnel operating on the scene or in the area providing support services. When the explosion occurred, one of the major initial problems was the accounting of all personnel on the site. This was difficult without an accountability system and added to confusion at the site. The duties of the safety officer include the management of risk on the fireground. Recognition of the oxygen-limiting silo's explosive potential and the minimal risk to property by the original situation would have fallen under the duties of the safety officer. The safety officer can also mitigate some of the burden on the incident commander and make that job less fatiguing.

6. **Unified command, which includes non-fire emergency managers, is essential for the successful resolution of a complex, long-term operation requiring extensive resources and generating a high level of media interest.**

According to dispatch and incident management protocols, the Emergency Director, EMS supervisor, the Fire Marshal, and the County Sheriff are notified of major incidents that may require response for coordination of resources and technical assistance. The Fire Marshal of Iredell County responded immediately after the explosion and stopped enroute to procure the services of a heavy construction crane. The early response of the crane was instrumental in stabilizing the rescue situation and was a factor in the successful removal of both trapped firefighters. The ability of the Emergency Management Director to coordinate non-fire resources was instrumental in incident operations as well.

7. **Technical rescue personnel from neighboring organizations should train together as a team as these operations are resource-intensive and require high maintenance of skill levels.**

At this incident, personnel from the various departments on the scene, as well as the Statesville units had successfully completed training courses beyond the basic technical rescue classes. The equipment, in the form of harnesses, line, hardware and the aerial tower, was of superior grade and type. However, the groups had not trained together nor adopted standard operational evolutions. Personnel interviewed indicated this was an issue and plans are underway to develop standard evolutions and train as a unified technical rescue effort across organizational boundaries.

8. **The use of personal protective clothing included the deployment of Class III harnesses prevented more serious injury and facilitated rescue.**

The use of full turnout gear including coats helmets and SCBA for the three firefighters on the silo roof prevented more serious injury from the explosion. While the anchor point for the harness attached to the three on the roof failed in the explosion, the use of harnesses facilitated rescue after the explosions. They eliminated the need for affixing additional knots and bets to rescue the injured firefighters. The harnesses performed as designed and were not damaged in the explosion.

APPENDIX A

Photographs

Photo 1. Silo with trailer shed to right and manufacturing facility in background

Appendix A (continued)

Photo 2. Access plate, origin of explosion

Appendix A (continued)

Photo 3. Access plate; area of origin of explosion

Appendix A (continued)

Photo 4. View of auger inside silo; fire burned along
the auger shafts to access panels

Appendix A (continued)

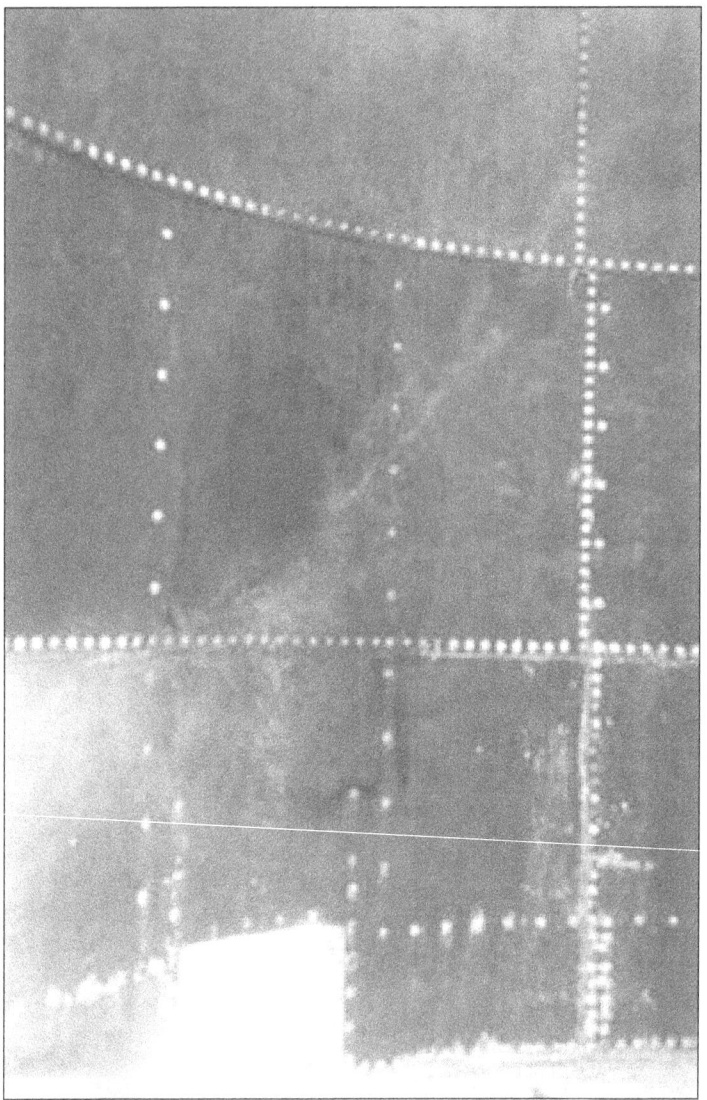

Photo 5. Smaller access panel indicating burning in
stack about 2' x 2' access panel (panel removed for
post-fire investigation)

Appendix A (continued)

Photo 6. Auger gear box, covered by a shield suspected origin of fire

Photo 7. Fire travel along auger shaft

Appendix A (continued)

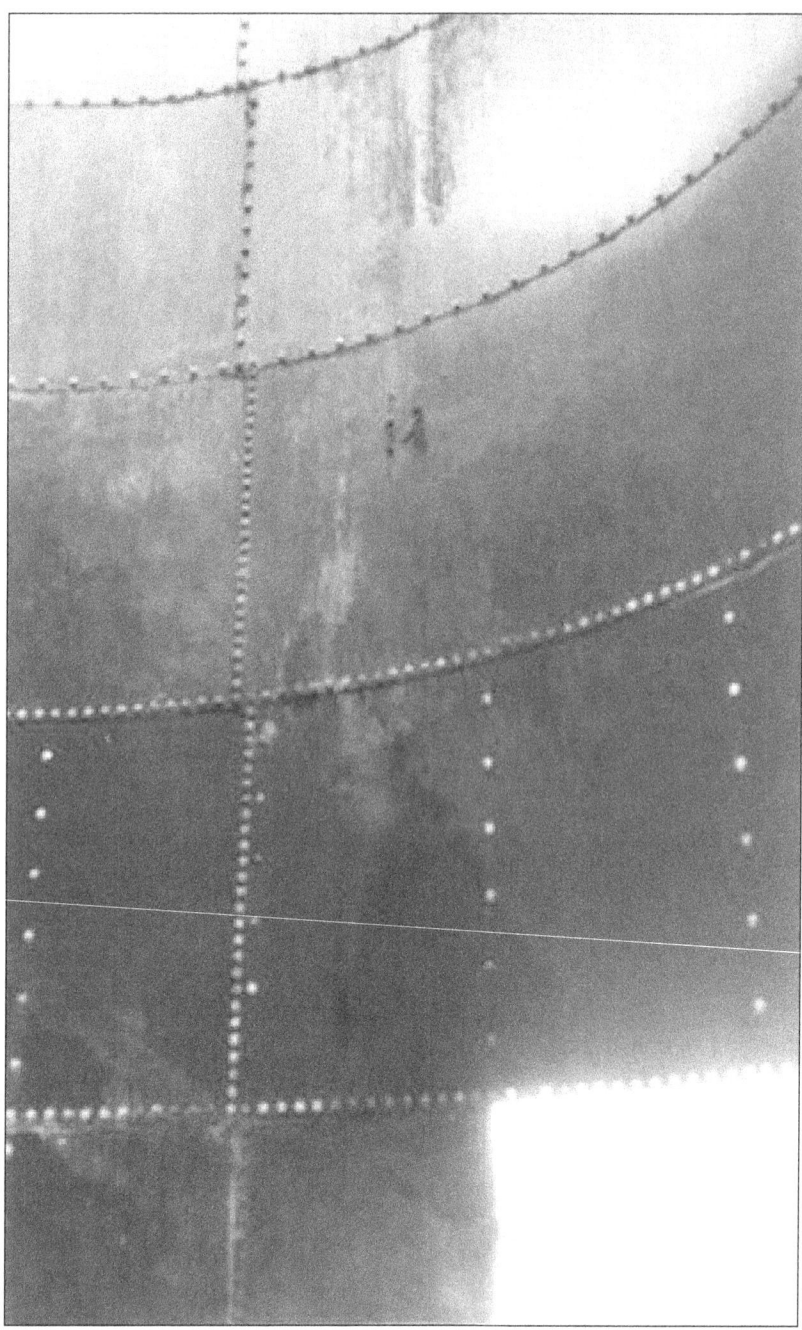

Photo 8. View from inside silo showing burned-out areas in
relation to access panel (origin of explosion)

APPENDIX B

Incident Chronology

1475 Winston Avenue, 12/21/97

Time of Dispatch: 08:58:49

Initial Alarm

0901:06-MVFD pumper	responding
0901:53-MVFD Chief	responding
0902:41-MVFD pumper	responding
0902:51-MVFD Chief	responding
0904:11-MVFD pumper	responding
0904:15-TVFD pumper	responding
0906:18-TVFD pumper	responding
0807:33-SIVFD pumper	responding

Support for Initial Alarm

0937:20-TVFD ambulance

0937:36-TVFD rescue

0952:36-MVFD support unit

1023:07-Air support unit

Approximate Time of Explosion: 11:40

Assistance after Explosion

1144:41-Medic 2	1509-Tower Ladder rescue completed, unit in service
1146:12-Tower Ladder, SFD	1711-Last fire unit leaves scene
1146:23-SFD Engine 2	
1153:01-Medic 3	MVFD-Monticello Volunteer Fire Department
1153:44-Medic 4	TVFD-Troutman Volunteer Fire Department
1153:51-Iredell ambulance	WIVFD-West Iredell Volunteer Fire Department
1154:31-Iredell ambulance	SFD-Statesville Fire Department
1156:37-SFD Engine 3	
1204:58-WIVFD Telesquirt	

www.ingramcontent.com/pod-product-compliance
Lightning Source LLC
Chambersburg PA
CBHW081251170526
45165CB00009B/3279